WORLD OF WONDER

Published by Creative Education
123 South Broad Street
Mankato, Minnesota 56001

Creative Education is an imprint of
The Creative Company.

Art direction by Rita Marshall
Design by The Design Lab
Editorial assistance by Will Koukkari and Kathleen Ball
Photographs by Corbis (Michael & Patricia Fogden), The Image Finders (Rob Curtis, Cynthia A. Delaney, Bill Leaman, Werner Lobert, Joanne Williams), Jay Ireland & Georgienne Bradley/Bradleyireland.com, JLM Visuals (D.F. & S.K. Austin, Don Blegen, J.C. Cokendolpher), Robert McCaw, James P. Rowan, Ann & Rob Simpson, Tom Stack & Associates (Brian Parker, Milton Rand, Dave Watts)

Library of Congress Cataloging-in-Publication Data

Hoff, Mary King.
Pollination / by Mary Hoff.
p. cm. – (World of wonder)
Summary: Describes how birds, insects, wind, and rain help plants reproduce by spreading their pollen.
ISBN 1-58341-270-0
1. Pollination–Juvenile literature. [1. Pollination.]
I. Title. II. World of wonder (Mankato, Minn.)

QK926 .H65 2003
571.8'642–dc21 2002035134

First Edition

9 8 7 6 5 4 3 2 1

cover & page 1: possums on eucalyptus
page 2: a dragonfly in search of food
page 3: a grasshopper spreading pollen

Creative Education presents

WORLD OF WONDER

POLLINATION

BY MARY HOFF

Flowers that look like bees ❦ Trees that fill the air with yellow dust ❦ Scents that lure bats ❄ The world is full of fascinating **adaptations** that enable seed-bearing plants to bring pollen grains (tiny, sperm-making particles produced by male parts of plants) in contact with egg-holding female parts of plants to make seeds.

THE PROCESS OF TRANSPORTING POLLEN grains from the male parts of plants to the female parts is called pollination. Pollination is one of the most important natural processes in the world, making it possible for seed-bearing plants to pass life on to another generation.

NATURE NOTE: *Pollen grains come in many sizes, shapes, and textures, from long and smooth to spherical and spiky.*

ON THE MOVE

Pollen grains produce sperm after they come in contact with the female part of a flower. The sperm carry the male part's contribution to seeds. In flowering plants, pollen grains are formed in a part of the flower called an **anther**. In **conifers**, they are produced by male cones. These particles are very tiny. A single plant may release millions of grains of pollen.

✿ In flowering plants, eggs are found in parts of the flower

The anthers of a tropical rain forest flower

These lily anthers hang under the flowers

called carpels. In conifers, eggs are found in female cones. Although many plants have both male and female parts—even in the same flower—in many cases the sperm and egg must come from different plants in order to produce a seed that will grow.

Reproduction is a big challenge for plants. In the animal world, males and females can

move around to find each other. But plants lack muscles and cannot run, fly, or crawl. In order to make seeds, they need something else to carry the pollen of one plant to the egg-bearing parts of another. Different plants use different pollen-carrying helpers. Some use the movement of air and water. Others rely on animals.

NATURE NOTE: *Even large pollen grains are smaller than the period at the end of this sentence.*

A water lily has both male and female parts

NATURE NOTE: *Most grasses are wind-pollinated. So are many trees, including most conifers, willows, and oaks.*

WIND AND WATER

In the summer, grasses in a field grow long stems. At the end of the stems are clusters of tiny flowers containing pollen. Even though these flowers aren't big and fancy like those of a rose or tulip, they are just as important in the plants' reproduction. When the wind blows, they release pollen grains, which float through the air. Some of these pollen grains land on the carpels of other grass plants, and the pollen-produced sperm and the egg together make a seed.

These milkweeds rely on help from the wind

Some plants use water instead of wind to move their pollen. One such plant is the northern or autumnal water-starwort. Its flowers release pollen underwater, and the water helps move the pollen to the female parts of other water-starwort flowers.

A plant called eel grass or wild celery, which lives in freshwater ponds, uses both water and wind in the pollination process. In this plant, the male and female parts are on separate flowers. The male flower,

Eel grass pollen rises from the water to the air

which develops underwater, produces little pollen-containing structures that break off and rise to the surface. They then float along until they bump into a female flower of the same species floating at the surface. When they bump, pollen falls onto the **stigma** of the female flower.

Rain sometimes plays a role in the pollination process,

NATURE NOTE: *The uncomfortable condition in humans called hay fever is an* **allergic reaction** *to pollen in the air.*

Rainwater can help unite pollen and eggs

too. Flowers of plants belonging to the buttercup family can be pollinated when rainwater carries pollen to the female parts of the plants.

NATURE NOTE: *The ragweed plant, which is wind-pollinated, can release a million pollen grains in a single day.*

Buttercups open wide to catch falling rain

INSECT TREATS

In plants that are pollinated by wind or water, only a tiny fraction of the pollen grains released end up on another plant. Some plants have a fascinating way of increasing the odds that their pollen will hit its target. Their flowers have features that attract certain insects. When an insect arrives, some of the plant's pollen gets

NATURE NOTE: *A kind of hawkmoth found on the island of Madagascar uses a proboscis that's about 12 inches (30 cm) long to reach nectar in flowers.*

stuck to its body. When the insect flies to another flower of the same species, it carries the pollen with it. Some of the pollen falls off onto the second flower's carpel, where it helps make a seed. Many different kinds of insects—including moths, butterflies, beetles, ants, flies, bees, and wasps—pollinate plants.

✲ In some plants, the pollen itself may be the attraction. Many insects, including beetles and honeybees, use pollen as a source of food. Other plants produce a sweet treat, called nectar, for their pollinators. Often the nectar is

Beetles and other insects seek pollen for food

A flower's bright colors can attract pollinators

NATURE NOTE: *Bees make honey by mixing nectar with special chemicals called enzymes in their mouths.*

located in the flower in such a way that insects have to brush against the pollen-packed anthers to get to it. Sometimes the nectar is stashed deep inside the flower. Insects that pollinate these flowers have very long feeding tubes, called proboscises, that make it possible for them to reach the nectar. Some plants give out other treats besides pollen and nectar. These include waxes, resins, and oils.

Plants that offer treats to pollinating insects "advertise" their offerings in various ways.

Some give off sweet scents. Many have bright colors that indicate the presence of food. Flowers that attract butterflies tend to be yellow or red, and flowers that attract bees tend to be yellow or blue. Some have **ultraviolet** patterns that act as road signs showing insects the way to the food they seek. Flowers that are pollinated by **nocturnal** moths are often light-colored and have strong fragrances to let pollinators know where they are in the dark.

Butterflies tend to favor yellow or red flowers

INSECT TRICKS

Some plants use tricks instead of treats to attract pollinating insects. Plants known as **carrion** plants have flowers that smell like rotting flesh. Some even look a little bit like rotting flesh

NATURE NOTE: *Some plants that lure insects by smelling like dead animals also create heat—just like the bodies of animals as they decompose.*

or like a bunch of flies feeding together. They attract insects that eat and lay their eggs on the dead bodies of other animals. When the insects search for the dead body that really isn't there, they collect pollen.

Plants fool insects in other ways, too. Some attract **predatory** insects by mimicking their prey. Others attract **parasitic** insects by looking like their hosts. For example, a type of orchid found in tropical habitats has flowers that look like spiders. They attract wasps that lay their eggs on spiders' bodies.

Skunk cabbage is named for its bad smell

* Another kind of trick attracts insects in search of a mate. Some orchid flowers look, smell, and feel like female bees. When a male bee tries to mate with the flower, he gets dusted with pollen. When he visits another flower of the same species, he deposits some of the pollen there.

BRING ON THE BIRDS

A red heliconia flower glows in the Costa Rican sun. A tiny rufous-tailed hummingbird flies up to it. The bird sticks its bill into the flower to suck up the nectar at its base. As it does, it picks up pollen. After drinking some nectar, the hummingbird flies on to another flower, carrying the pollen on its feathers.

NATURE NOTE: *Some bird-pollinated flowers have special structures that birds can use as perches while harvesting nectar.*

Pollinating birds tend to have long, thin bills

A bright flower refreshes a thirsty sunbird

⚜ The heliconia is just one example of plants that are pollinated by birds. Such plants can be found in many places around the world. In South Africa, sunbirds get food from the bright bird-of-paradise flower. In the South Pacific, brush-tongued parrots, or lories, pollinate flowering trees while gathering pollen and nectar for food. Other bird pollinators include

NATURE NOTE: *Red flowers are often pollinated by birds or butterflies, two kinds of animals that can see this color.*

sugarbirds and honeycreepers.

✤ Flowers that are pollinated by birds tend to be red or yellow and not strongly scented. These characteristics are related to birds' abilities to sense their surroundings. While they generally have very good eyesight, birds tend to have a poor sense of smell.

NATURE NOTE: *Hummingbirds' long tongues have tube-shaped grooves that help them draw nectar from inside flowers.*

Some pollinated plants produce delicious fruit

BAT ACTION

Some plants rely on bats as their pollination partners. Flowers that are pollinated by bats have special adaptations. They tend to be sturdy—an important trait, given their pollinators' relatively large size. They open in the evening or at night, when bats are active. They may be light-colored and have a strong scent that attracts their pollinators.

NATURE NOTE: *The flowers of a tropical vine called Mucana holtonii have bowl-shaped petals that allow bats to find them by bouncing sound waves off of them.*

Bat-pollinated plants often have a strong scent

Bats that eat nectar tend to have long noses and tongues. They collect pollen on their fur as they gather nectar from flowers.

Many plants found in tropical rain forests are pollinated by bats. Some cactus plants are also pollinated by bats. Bat-pollinated plants include wild papayas, bananas, mangoes, agave, saguaro cactuses, and organ pipe cactuses.

NATURE NOTE: *Many bats that pollinate flowers have brushy structures on the ends of their tongues to help them gather nectar and pollen.*

OTHER POLLINATORS

Although insects, birds, and bats are the most common pollination partners, they aren't the only ones. Proteas are plants found in southern Africa. Instead of holding their flowers high in the air where flying insects and birds can find them, some proteas have flowers that are hidden by their leaves and located near the ground. They give off a strong odor that attracts mice and rats. When the rodents root around in the plant, harvesting its rich nectar, they also gather pollen. When they feed on another protea flower, they pollinate the plant.

In Australia, the eucalyptus is pollinated by honey possums. These tiny mouse-like creatures have long tongues that allow them to gather nectar. They also have **prehensile** tails that allow them to cling to vegetation as they

NATURE NOTE: *Some trees pollinated by mammals have flowers growing right out of their trunks.*

feed. Geckos, a kind of lizard, pollinate shrubs in New Zealand. On the island of Madagascar, mammals called lemurs carry pollen as they move from plant to plant collecting nectar. Snails may also pollinate plants. As they crawl from one flower to another, they may carry pollen with them.

NATURE NOTE: *Scientists who study pollen are called palynologists. The word is related to the Greek word for "dust."*

Pollen collects on the fur of foraging lemurs

AMAZING ADAPTATIONS

From common grass flowers waving in the wind to elaborate blossoms that attract insects with sights or scents, plants use a variety of means to move pollen from one flower to another in order to reproduce. In almost all cases, plants must rely on help from either animals or the elements to carry out this process. ❄ These amazing adaptations are just a few examples of the many ways in which living things interact with each other in order to survive and thrive in the challenging world around them. They are valuable reminders of how intricately the lives of various creatures are intertwined. As humans make changes that affect the environment, it's important to remember and respect these connections. In doing so, we can help ensure the future health and beauty of this amazing world, this world of wonder.

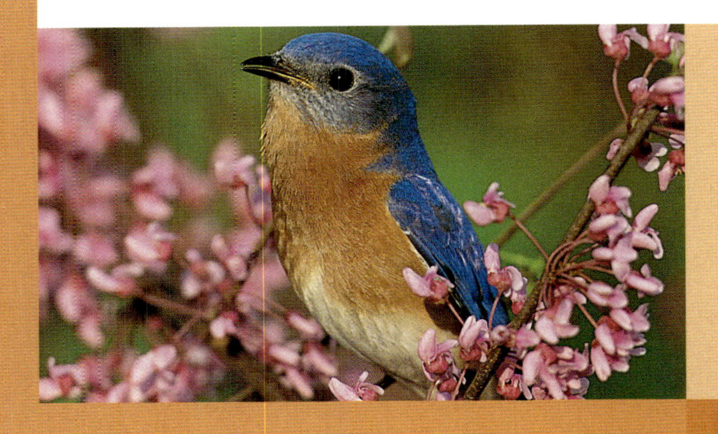

NATURE NOTE: *There are more than 1,000 known species of animals—other than insects—that pollinate plants.*

Pollination benefits both plants and animals

WORDS TO KNOW

Adaptations *are characteristics that contribute to a living thing's ability to survive or reproduce.*

An **allergic reaction** *is an unusually strong physical response (such as sneezing or itching) to certain substances.*

An **anther** *is a male structure in flowers that produces pollen.*

The remains of a dead animal are called **carrion**.

Conifers *are cone-bearing plants such as pines and spruce trees.*

Nocturnal *creatures are those that are active mainly at night.*

Parasitic *organisms draw energy and nutrients (and, in some cases, shelter) from other living organisms.*

Predatory *animals get the energy and nutrients they need by killing and eating other animals.*

A **prehensile** *tail is one that can grasp onto things such as a tree branch.*

A **stigma** *is a structure at the top of a plant's carpel where pollen lands and sticks.*

Ultraviolet *is a kind of light that can be seen by some insects but not by humans.*

INDEX